Juliana Berners, Wynkyn de Worde

A Treatise of Fysshynge wyth an Angle

Juliana Berners, Wynkyn de Worde

A Treatise of Fysshynge wyth an Angle

ISBN/EAN: 9783337250942

Printed in Europe, USA, Canada, Australia, Japan

Cover: Foto ©berggeist007 / pixelio.de

More available books at **www.hansebooks.com**

A Treatyse of

Fysshynge wyth an Angle

BY

DAME JULIANA BERNERS:

BEING A *FACSIMILE* REPRODUCTION OF THE FIRST BOOK ON THE

SUBJECT OF FISHING PRINTED IN ENGLAND BY

WYNKYN DE WORDE

AT WESTMINSTER IN 1496.

With an Introduction by

REV. M. G. WATKINS, M.A.

ELLIOT STOCK, 62, PATERNOSTER ROW,

LONDON, E.C.

1880

Preface

TO

Dame Juliana Berners' Treatyse on Fysshynge wyth an Angle.

THE fcholarly angler is here prefented with an exact *facfimile* of the firft Englifh treatife on fifhing. The book is of extreme intereft for feveral reafons, not the leaft curious being that it has ferved as a literary quarry to fo many fucceeding writers on fifhing, who have not difdained to adapt the authorefs's fentiments to their own ufe, and even to borrow them word for word without acknowledgment. Walton himfelf was evidently familiar with it, and has clearly taken his "jury of flies" from its "xij flyes wyth whyche ye fhall angle to ye trought & grayllyng;" while Burton, that univerfal plunderer, has extracted her eloquent eulogy on the fecondary pleafures of angling for incorporation with the patchwork structure of his "Anatomy of Melancholy." Befides giving the earlieft account of the art of fifhing, the eftimate which the authorefs forms of the moral value of the craft is not only very high, but has ferved to ftrike the keynote for all fubfequent followers of the art both in their praifes and their practice of it. To this little

treatife more than to any other belongs the credit of having
affigned in popular eftimation to the angler his meditative
and gentle nature.   Many pure and noble intellects have
kindled into lafting devotion to angling on reading her
eloquent commendation of it.   Such men as Donne, Wotton,
and Herbert, Paley, Bell, and Davy, together with many
another excellent and fimple difpofition, have caught enthufiafm from her lofty fentiments, and found that not their
bodily health only, but alfo their morals, were improved by
angling.   It became a fchool of virtues, a quiet paftime in
which, while looking into their own hearts, they learnt leffons
of the higheft wifdom, reverence, refignation, and love—
love of their fellow-men, of the lower creatures, and of their
Creator.

Nothing definite is known of the reputed authorefs, Dame
Juliana Barnes or Berners.   She is faid to have been a
daughter of Sir James Berners of Roding Berners in the
county of Effex, a favourite of King Richard the Second, who
was beheaded in 1388 as an evil counfellor to the king and
an enemy to the public weal.   She was celebrated for her
extreme beauty and great learning, and is reported to have
held the office of priorefs of the Benedictine Nunnery of
Sopwell in Hertfordfhire, a cell to the Abbey of St. Alban,
but of this no documentary evidence exifts.   The firft edition
of her "Book of St. Alban's," printed by the fchoolmafter-
printer of St. Alban's in 1486, treats of hawking, hunting,
and coat-armour.   In the next edition, "Enprynted at
Weftmeftre by Wynkyn the Worde the yere of thyncarnacõn
of our lorde. M.CCCC.lxxxxvi," among the other "treatyfes
perteynynge to hawkynge & huntynge with other dyuers
playfaunt materes belongynge vnto nobleffe," appeared the
prefent treatife on angling.   The ariftocratic inftincts of the

authoreſs prompted this mode of publication, as ſhe herſelf explains in the concluding paragraph—"by cauſe that this preſent treatyſe ſholde not come to the hondys of eche ydle perſone whyche wolde deſire it yf it were enprynted allone by itſelf & put in a lytyll plaunflet, therfore I haue com- pylyd it in a greter volume of dyuerſe bokys concernynge to gentyll & noble men to the entent that the forſayd ydle perſones whyche ſholde haue but lytyll meſure in the ſayd dyſporte of fyſſhyng ſholde not by this meane vtterly dyſtroye it." The preſent publication is the "little pamphlet" which was encloſed in this "greater volume." An edition of it as a diſtinct treatiſe appears to have been iſſued by Wynkyn de Worde ſoon after that of 1496, with the title, "Here be- gynnyth a treatyſe of fyſſhynge wyth an Angle" over the curious woodcut of the man fiſhing which is on the firſt page of the preſent *facſimile*, but only one copy of it is known to be in exiſtence. At leaſt ten more editions appeared before the year 1600. This ſhows the great popularity of the book at the time of its publication, and conſidering how human nature remains the ſame, and the charms of angling are equally grateful to every freſh generation of anglers, affords a ſufficient reaſon for the ſtrong antiquarian delight which all literary anglers of the preſent century have felt in the book. It is worth while briefly to trace the biblio- graphy of angling onwards until the appearance in 1653 of Walton's *Compleat Angler*, when the reader will be on familiar ground. In the interval of more than a hundred and fifty years between theſe two names of Berners and Walton, ſo deeply reverenced by every true ſcholar of the craft, there occur but four books on angling, though each one of theſe poſſeſſes a ſame peculiar to itſelf. Firſt came Leonard Maſcall's *Booke of Fiſhing with Hooke and Line*, publiſhed in

b

1590. Taverner's *Certaine Experiments concerning Fish and Fruite* followed in 1600. Then came in 1613 the *Secrets of Angling* of the celebrated angling poet, J. D. [John Dennys], whose verses have perhaps never yet been surpassed ; and finally, in 1651, appeared Barker's *Art of Angling*. With this fisherman and "ambassador's cook," as he calls himself, Walton must often have conversed.

It is a further testimony to the attractions which angling has always possessed for contemplative natures that the art appears here systematised, so to speak, as early as the middle of the fifteenth century in England, where it has been practised ever since with more enthusiasm and skill than in other countries. There is a sad gap in angling literature from the days of Ausonius, at the commencement of the fourth century, to those of Dame Juliana Berners. Fly-fishing, indeed, is not named between the time of Ælian and that of the Treatyse. It is clearly described by the former writer, who alone among the ancients mentions it, but in the present book it is spoken of under the term "angling with a dubbe," as if it were well-known and practised. Not only so, but it is clear that the writer had books of angling lore before her, perhaps monkish manuscripts, as Hawkins suggests, which would be of inestimable interest could they now be recovered. Thus in speaking of the carp, the reader will find she writes—"as touchynge his baytes I haue but lytyll knowlege of it. And me were loth to wryte more than I knowe & haue prouyd. But well I wote that the redde worme & the menow ben good baytys for hym at all tymes as I haue herde saye of persones credyble & also founde wryten in bokes of credence." No better rules can be given for fly-fishing at present than the two which she prescribes for angling—"for the fyrste and pryncypall poynt in anglynge : kepe ye euer fro the water fro

the fighte of the fyffhe," and "alfo loke that ye fhadow not
the water as moche as ye may." The "troughte" is to be
angled for "wyth a dubbe" [artificial fly] "in lepynge time;"
but as for the falmon, "ye may take hym: but it is feldom
feen with a dubbe at fuche tyme as whan he lepith in lyke
fourme & manere as ye doo take a troughte or a gryalynge."
With the imperfect tackle and clumfy rod of thofe days, it
is no wonder that the capture of falmon with a fly, which is
ftill the crowning achievement of the craft, could feldom be
effected.

After the eloquent pleading for angling with which the
treatife opens, the lady at once proceeds to teach the making
of the "harnays" of it. The rod fhe orders to be conftructed
fomewhat refembles, fave in its larger fize, the modern
walking-ftick rod. A hazel wand, or failing it, one of willow
or mountain afh, is to be procured, as thick as the arm and
nine feet in length. This is to form the butt, and is to be
hollowed out by means of divers red-hot irons into a taper-
ing hole, which is to receive the "croppe," or top, as we now
call it, when not in ufe. This "croppe" is to be made of
a yard of hazel, joined to a length of blackthorn, crab,
medlar, or "jenypre." All thefe are to be cut between
Michaelmas and Candlemas, the lady giving very particular
directions as to their drying and the like. When the two
portions of the "crop" are "fretted together," the whole
rod is to be fhaved into a fhapely taper form; the ftaff
encircled with long hoops of iron or latten at both ends,
and finifhed with a "pyke in the nether ende faftnyd wyth
a rennynge vyce: to take in & oute youre croppe." The line
is then to be wound round the crop and tied faft with a bow
at the top. The reader will note that there is no mention
of a reel; it was only ufed, feemingly until the beginning

of this century, for large falmon and pike. An angler who hooked a fifh when armed with this ponderous rod (which muft from its defcription have been nearly eighteen feet long, as large as a modern falmon rod), would act as Izaak Walton would have done in the like predicament,—throw the rod in to the fifh and recover it when he could. But the lady is wonderfully pleafed with this mighty rod, and thus concludes— "Thus fhall ye make you a rodde foo preuy that ye maye walke therwyth: and there fhall noo man wyte where abowte ye goo. It woll be lyghte & full nymbyll to fyffhe wyth at your lufte. And for the more redyneffe loo here a fygure," and fhe adds the curious woodcut which the reader may fee reproduced at page 5.

Then follow directions how to dye and make lines and hooks. There were evidently no manufacturers of hooks in the fifteenth century: each angler made his own. The cafting of plummets and forming of floats fucceed. The fix methods of angling and the mode of playing a fifh are next treated, and the latter alone fhows that Dame Juliana muft herfelf have been a proficient in the craft. No one but a thoroughly good fifher could have fummed up the art of playing a fifh in the words—"kepe hym euer vnder the rodde, and euermore holde hym ftreyghte: soo that your lyne may fufteyne and beere his lepys and his plungys wyth the helpe of your croppe & of your honde." The place, the time of day, and the weather in which to fifh, are next particularly defcribed after the exactitude peculiar to fifhing manuals of the olden time. Thefe paragraphs are well worth the confideration of a modern angler, efpecially the charge, "yf the wynde be in the Eeft, that is worfte For comynly neyther wynter nor fomer ye fyffhe woll not byte thenne."

The following part of the treatife, with what baits and how to angle for each kind of fifh, together with a brief defcription of each, certainly furnifhed Walton with a model for fome of his chapters. This portion of her book is regarded by the authorefs as moft neceffary to be known, and proficiency in carrying out her rules "is all the effecte of the crafte." She adds amufingly, "for ye can not brynge an hoke in to a fyffh mouth wythout a bayte." A few of the quaint receipts of her age succeed; how to keep live baits, to make paftes and the like, ending with a rule which is often given to flyfifhers for trout at the present day: "Whan ye haue take a grete fyffhe: vndo the mawe, & what ye fynde therin make that your bayte: for it is befte."

Juft as the authorefs rifes to eloquence at the beginning of the treatife when comparing the fifher's happy life with the toils and troubles which too often fall to the lot of the hunter, hawker, and fowler, so the end of thefe rules once more recalls her enthufiafm. The laft two pages of the book give us a portrait of her conception of the perfect angler, and it is no prefumption to fay that a nobler and truer picture has never been limned. Simplicity of difpofition, forbearance to our neighbours' rights, and confideration for the poor, are ftrongly inculcated. All covetoufnefs in fifhing or employment of its gentle art to increafe worldly gain and fill the larder is equally condemned. She holds the higheft view of angling; that it is to ferve a man for folace, and to caufe the health of his body, but efpecially of his soul. So fhe would have him purfue his craft alone for the moft part, when his mind can rife to high and holy things, and he may ferve God devoutly by faying from his heart his cuftomary prayer. Nor fhould a man ever carry his amufement to excefs, and catch too much at one time;

this is to deſtroy his future pleaſure and to interfere with that of his neighbours. A good ſportſman too, ſhe adds, will buſy himſelf in nouriſhing the game and deſtroying all vermin. So will what Walton calls "the civil, well-governed angler" eſcape the vices which ſpring from idle-neſs, and enjoy the full delights of an elevating and noble recreation. "And all thoſe that done after this rule ſhall haue the bleſſynge of god & ſaynt Petyr, whyche he theym graunte that wyth his precyous blood vs boughte."

"And therefore to al you that ben vertuous: gentyll: and free borne I wryte & make this symple treatyſe folow-ynge: by whyche ye may haue the full crafte of anglynge to dyſport you at your luſte: to the entent that your aege maye the more floure and the more lenger to endure."

M. G. W.

Salamon in his parablys sayth that a good spyryte makyth a flourynge aege/ that is a fayre aege & a longe. And syth it is soo: I aske this questyons. Whiche ben the meanes & the causes that enduce a man in to a mery spyryte.: Truly to my beste dyscrecōn it semeth good dysportes & honest gamys in whom a man Ioyeth wythout ony repentaunce after. Thenne folowyth it y goode dysportes & honest games ben cause of mannys fayr aege & longe life. And therfore now woll I chose of foure good dysportes & honeste gamys/ that is to wyte: of huntynge: hawkynge: fysshynge: & foulynge. The beste to my symple dyscrecōn why the is fysshynge: callyd Anglynge wyth a rodde : and a lyne

and an hokes. And therof to treate as my symple wytte may suf
fyce:both for the sayd reason of Salamon and also for the rea
son that phisyk maketh in this wyse(¶ Si tibi deficiant medici
medici tibi fiant:hec tria mens leta labor & moderata dieta.
¶ Ye shall vnderstonde that this is for to saye / yf a man lacke
leche or medicyne he shall make thre thynges his leche & medy
cyne:and he shall nede neuer no moo. The fyrste of chepm is a
mery thought.The seconde is labour not outrageod?.The thyr
de is dyete mesurable . Fyrste that yf a man wyll euer more be
in mery thoughtes and haue a glande spyryte:he must eschewe
all contraryous company & all places of debate where he my =
ghte haue ony occasyons of malencoly . And yf he woll haue a
labour not outrageous he must thenne ordeyne him to his her
tys ease and pleasaunce wythout studye pensyfnesse or trauey
le a mery occupacyon whyche maye reioyce his herte:& in why
che his spyrytes may haue a mery delyte.And yf he woll be dy
etyd mesurably he must eschewe all places of ryotte whyche is
cause of surfette and of syknesses. And he must drawe him to pla
ces of swete ayre and hungry:And ete nourishable meetes and
dyffyable also.

NOw thenne woll I dyscryue the sayd dysportes and ga
mys to fynde the beste of chepm as veryly as I can.sall
be it that the ryght noble and full worthy prynce the du
ke of Yorke late callid mayster of game hath dyscryued the myr
thes of huntynge lyke as I thynke to dyscryue of it and of alle
the other.For huntynge as to myn entent is to laboryous/For
the hunter must alwaye renne & folowe his houndes : trauey-
lynge & swetynge full sore.He blowyth tyll his lyppes blyster
And whan he wenyth it be an hare full oft it is an hegge hogge
Thus he chasyth and wote not what.He comyth home at euyn
rayn beteynyrycleyd:and his clothes torne were shode all myry
Some hounde loste:some surbat.Suche greues & many other
happyth vnto the hunter / whyche for dyspleysaunce of chepm y
loue it I dare not reporte . Thus trulye me semyth that this is
not the beste dysporte and game of the sayd foure.The dyspor
te and game of hawkynge is laboryous & noyouse also as me
semyth.For often the fawkener leseth his hawkes as the hun-

ter his houndes. Thenne is his game & his dysporte goon. Full
often aspeth he & whystelpth tyll that he be ryght euyll a thur
ste. Hys hawke taketh a bowe and lyste not ones on hym rewar
de. Whan he wolde haue her for to flee: thenne woll she bathe .
With mys feodynge she shall haue the Fronse: the Rpe: the Cray
and many other sykneſſes that brynge theym to the Sowle.
Thus by prouff this is not the beſte dysporte & game of the ſa
pd foure. The dysporte & game of fowlynge me ſempth mooſt
ſymple For in the wynter ſeaſon the fowler ſpeodyth not but in
the mooſt hardeſt and coldeſt weder: whpche is greuous . For
whan he wolde goo to his gynnes he mape not for colde. Ma=
ny a gynne & many a ſnare he makyth. Yet ſorylp dooth he fa=
re. At morn tyde in the dewe he is weete shode vnto his taylle.
Many other ſuche I cowde tell: but drede of magre makith me
for to leue. Thus me ſempth that huntynge & hawkynge & al=
ſo fowlynge ben ſo laborous and greuous that none of theym
mape perfourme nor bi verp meane that enduce a man to a me
ry ſpyrpte: whpche is cauſe of his longe lyfe acordynge vnto þ
ſayd parable of Salamon. ¶Dowteles theñe folowyth it that
it muſt nedes be the dysporte of fyſſhynge wpth an angle. For
all other manere of fyſſhyng is alſo laborous & greuous: often
makynge folkes ful wete & colde) whpche many tymes hath be
ſeen cauſe of greete Infirmytees. But the angler mape haue no
colde nor no dyſeaſe nor angres but yf he be cauſer hymſelf. For
he mape not leſe at the mooſt but a lyne or an hoke: of whpche
he mape haue ſtore plentee of his owne makynge, as this ſym
ple treatyſe shall teche hym. Soo thenne his loſſe is not greuo
us . and other grepffes mape he not haue) ſaupnge but yf ony
fiſſhe breke away after that he is take on the hoke, or elles that
he catche nought : whpche ben not greuous . For yf he faylle of
one he mape not faylle of a nother) yf he dooth as this treaty=
ſe techyth: but yf there be nonght in the water. And yet atte the
leeſt he hath his holſom walke and mery at his eaſe. a ſwete ay
re of the ſwete ſauoure of the meede floures: that makyth hym
hungry. He hereth the melodyous armony of fowles. He ſeeth
the ponge Swannes: heerons: duckes: cotes and many other fou
les wpth thepr brodes.) whpche me ſempth better than alle the

noyse of honndys:the blastes of hornys and the scrye of foulis
that hunters:fawkeners & foulers can make. And yf the angler
take fysshe:surely thenne is there noo man merier than he is in
his spyryte. ¶ Also who soo woll vse the game of anglynge :he
must ryse erly.whiche thyng is prouffytable to man in this wy
se.That is to wyte:moost to the heele of his soule. For it shall
cause hym to be holp.and to the heele of his body. For it shall
cause hym to be hole. Also to the encreale of his gooddys . For it
shall make hym ryche. As the olde englysshe prouerbe sayth in
this wyse.¶ Who soo woll ryse erly shall be holp helthy & zely.
¶ Thus haue I proupd in myn entent that the dysporte & ga=
me of anglynge is the very meane & cause that endurith a man
in to a mery spyryte:Whyche after the sayde parable of Salo-
mon & the sayd doctryne of phisyk makyth a flourynge aege &
a longe. And therfore to al you that ben vertuous:gentyll:and
free borne I wryte & make this symple treatyse folowynge: by
whyche ye may haue the full crafte of anglynge to dysport you
at your luste:to the entent that your aege maye the more flou
re and the more lenger to endure.

YF ye woll be crafty in anglynge : ye must fyrste lerne to
make your harnays. That is to wyte your rodde:your
lynes of dyuers colours. After that ye must know how
ye shall angle in what place of the water:how depe:and what ti
me of day. For what manere of fysshe:in what wedyr. How ma
ny impedymentes there ben in fysshynge y is callpd anglynge
And in specyall wyth what baytys to euery dyuers fysshe in e=
che moneth of the yere. How ye shall make your baytes brede
Where ye shall fynde thepm:and how ye shall kepe thepm . And
for the moost crafty thynge how ye shall make youre hokes of
stele & of osmonde. Some for the dubbe:and some for the flote:
& the grounde.as ye shall here after al thyse fynde expressed o=
penly vnto your knowlege.
¶ And how ye shall make your rodde craftly here I shall teche
you.Ye shall kytte betwene Myghelmas & Candylmas a fayr
staffe of a fadom and an halfe longe:& arme grete of haspll:wy
lowe:or aspe. And bethe hym in an hote ouyn:& sette hym euyn
Thenne lete hym cole & drye a moneth . Take thenne & frette

hym faste wyth a cockeshotecorde; and bynde hym to a fourme
or ag.euyn square grete tree.Take thenne a plumers wire that
is euyn and strepte & sharpe at the one ende.And hete the shar
pe ende iŋ a charcole fyre tyll it be whyte:and brenne the staffe
therwyth thorugh:euer strepte iŋ the ppthe at bothe endes tyll
they mete. And after that brenne hym iŋ the nether ende wyth
a byrde broche:& wyth other broches eche gretter than other.&
euer the grettest the laste : so that pe make pour hole aye tapre
were. Thenne lete hym lye styll and kele two dayes . Vnfrette
hym thenne.and ler e hym drye iŋ aŋ hous roof iŋ the smoke tyll
he be thrugh drye ¶ Iŋ the same seasoŋ take a fayr perde ofgre
ne haspll & beth hym euyn & strepghte.and lete it drye with the
staffe . And whan they bey drye make the perde mete vnto the
hole iŋ the staffe:vnto halfe the length of the staffe. And to per
fourme that other halfe of the croppe.Take a fayr shote of blac
ke thoriŋ:crabbe tree:medeler.or of Jenypre hytte iŋ the same se
asoŋ:and well bethyd & strepghte.And frette theym togyder fe
tely:soo that the croppe maye iustly entre all iŋ to the sayd ho=
le.Thenne shaue pour staffe & make hym tapre were. Thenne
vprell the staffe at bothe endes wyth longe hopis of prey or la
toŋ iŋ the clenŋest wile wyth a ppke iŋ the nether ende faltnyd
wyth a rennynge vyce:to take iŋ & oute poure croppe.Thenne
fet pour croppe aŋ handfull within the ouer ende of pour staffe
iŋ suche wile that it be as bigge there as iŋ oŋy other place abo
ue . Theñe arme pour croppe at thouer ende dowŋe to ẏ frette
wyth a lyne of.vj.heeres.And dubbe the lyne and frette it faſt
iŋ ẏ toppe wyth a bowe to faſteŋ oŋ pour lyne . And thus shall
pe make pou a rodde soo preuy that pe maye walke therwyth :
and there shall noo maŋ wyte where abowte pe goo. It woll be
leghte & full nymbyll to fyſhe wyth at pour lufte. And for the
more redynelle loo here a fygure therof iŋ example.:

Fter that pe haue made thus pour rodde: pe muſt lerne
to coloure pour lynes of here iŋ this wyle . ¶Fprſte pe
muſt take of a whyte horſe taylle the lengeſt heere and

faprest that ye can fynde. And euer the rounder it be the better
it is. Departe it in to .vj. partes: and euery parte ye shal colour
by hymselfe in dyuers colours. As yelowe: grene: browne: taw=
ney: russet. and duske colours. And for to make a good grene co
lour on your heer ye shall doo thus. ¶ Take smalle ale a quar
te and put it in a lytyll panne: and put therto halfe a pounde of
alym. And put therto your heer: and lete it boylle softly half an
houre. Thenne take out your heer and lete it drye. Thenne ta
ke a potell of water and put it in a panne . And put therin two
handfull of oolops or of wyren . And presse it wyth a tyle sto=
ne: and lete it boylle softly half an houre. And whan it is yelow
on the scume put therin your heer wyth halfe a pounde of copo
rose betyn in powdre and lete it boylle halfe a myle wape: and
thenne sette it downe: and lete it kele fyue or syxe houres. Then
take out the heer and drye it. And it is thenne the fynest grene
that is for the water. And euer the more ye put therto of copo
rose the better it is. or elles in stede of it vertgrees.
¶ A nother wyse ye maye make more bryghter grene as thus
Lete woode your heer in an woodefatte a lyght plunket colour
And thenne sethe hym in olde or wyryn lyke as I haue sayd: sa=
uynge ye shall not put therto neyther coporose ne vertgrees.
¶ For to make your heer yelow dyght it wyth alym as I haue
sayd before. And after that wyth oldys or wyryn wythout copo
rose or vertgrees.   ¶ A nother yelow ye shal make thus. Ta
ke smalle ale a potell: and stampe thre handful of walnot leues
and put togider: And put in your heer tyll that it be as depe as
ye woll haue it.   ¶ For to make russet heer. Take stronge lye
a pynt and halfe a pounde of sote and a lytyll iuce of walnot le
ups a quarte of alym: and put thepm alle togyder in a panne
and boylle thepm well . And whan it is colde put in youre heer
tyll it be as derke as ye woll haue it.   ¶ For to make a brow
ne colour. Take a pounde of sote and a quarte of ale: and seth it
wyth as many walnot leups as ye maye. And whan they were
blacke sette it from the fyre. And put therin your heer and lete it
lye styll tyll it be as browne as ye woll haue it.
¶ For to make a nother browne. Take strong ale and sote and
tempre them togyder. and put therin your heer two dayes and
two nyghtes and it shall be ryght a good colour.

⸿For to make a tawney coloure. Take lyme and water ⁊ put thepm togyder: and also put pour heer therin foure oꝛ fyue hou res. Thenne take it out and put it in a Tanners ofe a dap: and it ſhall be also fyne a tawney colour as nedyth to our purpoos ⸿The fyrſte parte of pour heer ye ſhall kepe ſtyll whyte foꝛ ly nes foꝛ the dubbyd hoke to fyſſhe foꝛ the tꝛought and graplyn ge. and foꝛ ſmalle lynes foꝛ to rye foꝛ the roche and the darſe

Whan pour heer is thus colourid: ye muſt knowe foꝛ whi che waters and foꝛ whpche ſeaſons thep ſhall ſerue.
⸿The grene colour in all cleꝛe water from Apꝛyll tyll Septembre. ⸿The yelowe coloure in euery cleꝛe water from Septembre tyll Nouembre: Foꝛ is is lyke y̆ wedys and other mánere graſſe whiche gꝛowyth in the waters and ryuers whan thep ben broken. ⸿The ruſſet colour ſerupth all the wynter vnto the ende of Apꝛyll las well in ryuers as in poles oꝛ lakys ⸿The browne colour ſerupth foꝛ that water that is blacke de diſſhe in ryuers oꝛ in other waters. ⸿The tawney colour foꝛ thoſe waters that ben hethy oꝛ moꝛyſſhe.

Now muſt ye make poure lynes in this wyſe. Fyrſte lo= ke that ye haue an Jnſtꝛument lyke vnto this fygure poꝛtꝛayed folowynge. Thenne take pour heer ⁊ kytte of the ſmalle ende an hondfull large oꝛ moꝛe: foꝛ it is neyther ſtꝛonge noꝛ pet ſure. Thenne toꝛne the toppe to the taylle eue rytche plyke moche. And departe it into thꝛe partyes. Thenne knytte euery part at the one ende by hymſelf. And at the other ende knytte all thꝛe togyder: and put y̆ ſame ende in that other ende of pour Jnſtꝛument that hath but one clyſt. And ſett that other ende faſte wyth the wegge foure fyngers in alle ſhoꝛter than pour heer. Thenne twyne euery warpe one waye ⁊ plyke moche: and faſten thepm in thꝛe clyſtes plyke ſtrepghte. Take thenne out that other ende and twyne it that waye that it woll deſyre ynough. Thenne ſtꝛepne it a lytyll: and knytte it foꝛ vn dopnge: and that is good. And foꝛ to knowe to make pour Jn= ſtꝛument: loo heꝛe it is in fygure. And it ſhall be made of tꝛee ſawynge the bolte vndeꝛneth: whiche ſhall be of yꝛen.

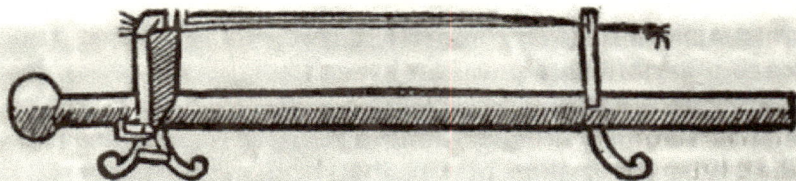

WHan þe haue as many of the lynkys as ye suppose wol
suffyse for the length of a lyne :thenne must ye knytte
theym togyder wyth a water knotte or elles a duchys
knotte. And whan your knotte is knytte:kytte of ȳ voyde shor
te endes a shalbe brede for the knotte. Thus shal ye make you
re lynes fayr & fyne:and also ryght sure for ony manere fysshe.
¶And by cause that ye sholde knowe bothe the water knotte &
also the duchys knotte:loo theym here in fygure caste vnto the
lyknesse of the draughte.

YE shall vnderstonde that the moost subtyll & hardyste
crafte in makynge of your harnays is for to make your
hokis. For whoos makyng ye must haue fete fyles.thyn
and sharpe & smalle beten:A semy clam of pyrey:a bender:a pa
yr of longe & smalle tongys : an harde knyfe somdeale thycke:
an anuelde:& a lytyll hamour. ¶And for smalle fysshe ye shall
make your hokes of the smalest quarell nedlys that ye can fyn
de of stele & in this wyse. ¶Ye shall put the quarell in a redde
charkcole fyre tyll that it be of the same colour that the fyre is.
Thenne take hym out and lete hym kele:and ye shal fynde him
well alayd for to fyle. Thenne reyse the berde wyth your kny
fe:and make the poynt sharpe. Thenne alaye hym agayn: for
elles he woll breke in the bendyng. Thenne bende hym lyke to
the bende fyguryd herafter in example. And gretter hokes ye
shall make iȝ the same wyse of gretter nedles:as brodyers ne
dlis:or taylers:or shomakers nedlis spere poyntes &

h iij

of ſhomakers nalles in eſpecpall the beſte for grete fpſſhe . and
that they bende atte the popnt whan they ben aſſapedſfor elles
they ben not good ¶ whan the hoke is bendpd bete the hynder
ende abrode: ⁊ fyle it ſmothe for fretynge of thy lyne. Thenne
put it in the fyre agapn: and peue it an eaſy redde here. Thenne
ſodaynlp quenche it in water: and it woll be harde ⁊ ſtronge.
And for to haue knowlege of pour Inſtrumentes: lo thepm he=
re in fygure portrapd.

¶ Hamour.   Knyfe.   Pynſons.   Claſſ

Wegge.   Fyle.   Wreſte.   ⁊ Anuelde.

Whan pe haue made thus pour hokis: thenne muſt pe ſet
thepm on pour lynes acordynge in gretneſſe ⁊ ſtrength
in this wyſe. ¶ Pe ſhall take ſmalle redde ſilke. ⁊ pf it be
for a grete hoke theñe double it: not twynyd. And elles for ſma
le hokps lete it be ſpngle: ⁊ therwpth frette thpcke the lyne the
re as the one ende of pour hoke ſhal ſytte a ſtrawe brede. Theñ
ſette there pour hoke: ⁊ frette hpm wpth the ſame threde p two
partes of the lengthe that ſhall be frette in all. And whan pe co
me to the thprde part e thenne torne the ende of pour lyne aga
pn vpon the frette dowble. ⁊ frette it ſo dowble that other thyr
de part e. Thenne put pour threde in at the hole twys or thries
⁊ lete it goo at eche tyme rounde abowte the perde of pour ho=
ke. Thenne wete the hole ⁊ drawe it tyll that it be faſte. And lo
ke that pour lyne lye euermore wpthin pour hokps: ⁊ not with
out. Thenne kptte of the lynps ende ⁊ the threde as nyghe as
pe mape: ſaupnge the frette.

Now pe knowe wpth how grete hokps pe ſhall angle to
euerp fpſſhe: now I woll tell pou wpth how manp hee=
res pe ſhall to euerp manere of fiſſhe. ¶ For the menow
wpth a lyne of one heere. For the warpng roche. the blcke ⁊ the

gogyn ⁊ the ruffe wyt a lyne of two heeris. For the darfe ⁊ the
grete roche wyth a lyne of thre heeres. For the perche:the flou
der ⁊ bremet with foure heeres. For the cheuen chubbe:the bre
me:the tenche ⁊ the ele wyth. vj. heeres. For the troughte:grap
lynge:barbyll ⁊ the grete cheupy wyth. ix. heeres. For the grete
troughte wyth. xij. heeres: For the famon wyth. xv. heeres. And
for the ppke wyth a chalke lyne made browne with pour brow
ne colour aforfayd: armyd with a wyre. as pe fhal here herafter
whan I fpeke of the ppke.

Pour lynes muft be plumbid wyth lede. And pe fhall wyte þ
the nexte pūbe vnto the hoke fhall be therfro a large fote ⁊ mo
re. And euery plumbe of a quantyte to the gretnes of the lyne.
There be thre manere of plūbis for a grounde lyne rennynge.
And for the flote fet vpon the grounde lyne lyenge. x. plumbes
Ioynynge all togider. On the grounde lyne rennynge. ix or. x.
fmalle. The flote plūbe fhall be fo heup þ the leeft plucke of o=
ny fyfh he mape pull it downe in to ꝩ water. And make pour plū
bis rounde ⁊ fmothe þ they ftycke not on ftonys or on wedys.
And for the more vnderftondynge lo theym here in fygure.

The grounde lyne rennynge

The grounde lyne lyenge.

The flote lyne

The lyne for perche or tenche.

The lyne for a ppke: Plūbe: Corke armyd wyth wyre

Thenne fhall pe make pour flotys in this wyfe. Take a
fayr corke that is clene without many holes. and bore it

thrugh wyth a smalle hote pren: And putt therin a penne luste
and straghte. Euer the more flote the gretter penne & the gre
ter hole. Thenne shape it grete in the mydois and smalle at bo
the endys. and specyally sharpe in the nether ende⸗ and lyke vn
to the fygures folowynge. And make thepm smothe on a gryn
dyng stone: or on a tyle stone. ❡ And loke that the flote for one
heer be nomore than a pese. For two heeres: as a beene. for twel
ue heeres: as a walnot. And soo euery lyne after the proporcōg,
❡ All manere lynes that bey not for the groude must haue flo
tes. And the rennynge groude lyne must haue a flote. The ly
enge grounde lyne wythout flote.

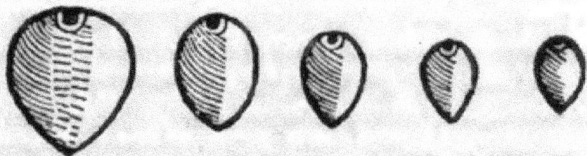

NOw J haue lernyd you to make all your harnays. He⸗
re J woll tell you how ye shall angle. ❡ Ye shall angle:
vnderstonde that there is. vj. manere of anglyng. That
one is at the grounde for the troughte and other fisshe. A no⸗
ther is at p̌ grounde at an archefor at a stange where it ebbyth
and flowyth: for bleke: roche. and darse. The thyrde is wyth a
flote for all manere of fysshe. The fourth wyth a menow for y̌
troughte wythout plumbe or flote. The fyfth is rennynge in y̌
same wyse for roche and darse wyth one or two heeres & a flye.
The syxte is wyth a dubbyd hoke for the troughte & graplyng
❡ And for the fyrste and pryncypall poynt in anglynge: kepe y̌
euer fro the water fro the sighte of the fysshe: other ferre on the
londe: or ellys behynde a busshe that the fysshe se you not . For
yf they doo they wol not byte. ❡ Also loke that ye shadow not
the water as moche as ye may. For it is that thynge that woll
soone frape the fysshe. And yf a fysshe be afrayed he woll not bi
te longe after. For alle manere fysshe that fede by the grounde
ye shall angle for theim to the botom. soo that your hokys shall
renne or lye on the grounde. And for alle other fysshe that fede

aboue þe shall angle to theym in the myddes of the water or
somdeale byneth or somdeale aboue. For euer the gretter fisshe
the nerer he lyeth the botom of the water. And euer the smaller
fysshe the more he swymmyth aboue. ¶The thyrde good poynt is whan the fysshe bytyth that ye be not to hasty to smyte
nor to late. For ye must abide tyll ye suppose that the bayte be
ferre in the mouth of the fysshe, and thenne abyde noo longer.
And this is for the grounde. ¶And for the flote whan ye se it pul
lyd softly vnder the water: or elles caryed vpon the water softly: thenne smyte. And loke that ye neuer ouersmyte the strengthe of your lyne for brekynge. ¶And yf it fortune you to smyte a grete fysshe wyth a smalle harnays: thenne ye must lede
hym in the water and labour him there tyll he be drowynd and
ouercome. Thenne take hym as well as ye can or maye, and e-
uer bewaar that ye holde not ouer the strengthe of your lyne.
And as moche as ye may lete hym not come out of your lynes
ende streyghte from you: But kepe hym euer vnder the rodde,
and euermore holde hym streyghte: soo that your lyne may sus
teyne and beere his leppys and his plungys wyth the helpe of
your croppe & of your honde.

Here I woll declare vnto you in what place of the water
ye shall angle. Ye shall angle in a pole or in a stondinge
water in euery place where it is ony thynge depe. The
re is not grete choyse of ony places where it is ony thynge de
pe in a pole. For it is but a pryson to fysshe. and they lyue for ye
more parte in hungre lyke prisoners: and therfore it is the lesse
maystry to take theym. But in a ryuer ye shall angle in euery
place where it is depe and clere by the grounde: as grauell or
claye wythout mudde or wedys. And in especyall yf that there
be a manere whyrlynge of water or a couert. As an holow ban
ke: or grete rotys of trees: or longe wedes fletyng aboue in the
water where the fysshe maye couere and hyde theymself at cer-
tayn tymes whan they lyste Also it is good for to angle in depe styffe stremys and also in fallys of waters and weares: and
in floode gatys and mylle pyttes. And it is good for to angle
where as the water restyth by the banke: and where the streme
rennyth nyghe there by: and is depe and clere by the grounde

and in ony other placys where ye may se ony fyssh houe or ha
ne ony seodynge.

Ow ye shall wryte what tyme of the daye ye shall angle
¶From the begynnynge of May vntyll it be Septem
bre the bytynge tyme is erly by the morowe from fou
re o fy cloche vnto eyghte of the cloche. And at after none from
foure of the clocke vnto eyghte of the cloche : but not soo good
as is in the mornynge. And yf it be a colde whystelyng wynde
and a derke lowrynge daye . For a derke daye is moche better
to angle in than a clere daye. ¶From the begynnynge of Sep
rembre vnto the ende of Apryll spare noo tyme of the daye:
¶Also many pole fysshes woll byte beste in the none tyde.
¶And yf ye se ony tyme of the daye the trought or graylynge
lepe: angle to hym wyth a dubbe accordynge to the same month
And where the water ebbyth and flowyth the fysshe woll byte
in some place at the ebbe: and in some place at the flood. After þ
they haue restynge behynde stangnys and archys of brydgys
and other suche manere places.

Ere ye shall wryte in what weder ye shall angle. as I sa
yd before in a derke lowrynge daye whanne the wynde
blowyth softly. And in somer season whan it is brennyn
ge hote thenne it is nought . ¶From Septembre vnto Apryll
in a fayr sonny daye is ryght good to angle. And yf the wynde
in that season haue ony parte of the Oryent: the wedyr thenne
is nought. And whan it is a grete wynde. And whan it snowieh
ceynyth or haylyth. or is a grete tempeste as thondyr or ligh
tenynge: or a swoly hote weder: thenne it is noughte for to an=
gle.

Ow shall ye wryte that there ben twelue manere of ympi=
pedymentes whyche cause a may to take noo fysshe. wt
out other compy that maye casuelly happe. ¶The fyrst
is yf your harnays be not mete noo fetly made. The seconde is
yf your baytes be not good nor fyne. The thyrde is yf that ye
angle not in bytynge tyme. The fourth is yf that the fysshe be
frayed wt the syghte of a may. The fyfth yf the water be very
thycke: whyte or redde of ony floode late fallen. The syxte yf
the fysshe styre not for colde . The seuenth yf that the wedyr

be hote. The eyght yf it rayne. The nynthe yf it hayll or snow̄ falle. The tenth is yf it be a tempeste. The enleuenth is yf it be a grete wynde. The twelfyth yf the wynde be in the Eest and that is worste. For comenly neyther wynter nor somer ȳ fysshe woll not byte thenne. The weste and northe wyndes ben good but the south is beste.

ANd now I haue tolde you how to make your harnays: and how ye shall fysshe therwyth in al poyntes Reason woll that ye knowe wyth what baytes ye shall angle to euery manere of fysshe in euery moneth of the yere ƺ whyche is all the effecte of the crafte. And wythout whyche baytes know en well by you all your other crafte here toforn auayllyth you not to purpose. For ye can not brynge an hoke in to a fyssh mo uth wythout a bayte. Whiche baytes for euery manere of fyssh and for euery moneth here folowyth in this wyse.

FOr by cause that the Samon is the moost stately fyssh that ony man maye angle to in fresshe water. Therfore I purpose to begyn at hym . ¶The samon is a gentyll fysshe: but he is comborous for to take . For comenly he is but in depe places of grete ryuers. And for the more parte he hol dyth the myddys of it: that a man maye not come at hym. And he is in season from Marche vnto Myghelmas . ¶In whyche season ye shall angle to hym wyth thyse baytes whan ye maye gete thepm. Fyrste wyth a redde worme in the begynnynge ƺ endynge of the season. And also wyth a bobbe that bredyth in ȝ dunghyll. And specyally wyth a souerayn bayte that bredyth on a water docke. ¶And he byteth not at the grounde: but at ȳ flote. Also ye may take hym : but it is seldom seen with a dubbe at suche tyme as whan he lepith in lyke fourme ƺ maners as ye doo take a troughte or a grayfynge. And thyse baytes ben well proued baytes for the samon.

THe Troughte for by cause he is a right deyntous fyssh and also a ryght feruente-byter we shall speke. nexte of hym. he is in season fro Marche vnto Myghelmas. He is on clene grauely groū de ƺ in a streme. Ye may angle to hym

all tymes wyth a grounde lyne lyenge oꝛ rennynge: sauyng iꝝ
leppynge tyme.and thenne wyth a dubbe. And eꝛly wyth a ren꙯
nynge grounde lyne.and foꝛth iꝝ the dayꝫ wyth a flote lyne.
℧ He ſhall angle to hym iꝝ Marche wyth a menew hangyd on
pour hoke by the nether neſſe wythout flote oꝛ plumbe:draw꙯
pnge vp ⁊ downe iꝝ the ſtꝛeme tyll pe fele hym taſte. ℧ Iꝝ the
ſame tyme angle to hym wyth a groūde lyne with a redde woꝛ
me foꝛ the mooſt ſure. ℧ Iꝝ Apꝛill take the ſame baytes: ⁊ alſo
Inneba other wyſe nampd.vij. eyes. Alſo the canker that bre꙯
dyth iꝝ a grete tꝛee and the redde ſnayll. ℧ Iꝝ May take y ſto
ne flye and the bobbe vnder the cowe toꝛde and the ſylke woꝛ꙯
me: and the bayte that bredyth on a ferꝝ lepf. ℧ Iꝝ Iuyꝝ take a
redde woꝛme ⁊ nyppe of the heed: and put on thyꝝ hoke a cod꙯
woꝛme byfoꝛꝝ. ℧ Iꝝ Iulll take the grete redde woꝛme and the
codwoꝛme togyder. ℧ Iꝝ Auguſt take a fleſſhe flye ⁊ the grete
redde woꝛme and the fatte of the bakoꝝ: and bynde abowte thy
hoke. ℧ Iꝝ Septembre take the redde woꝛme and the menew.
℧ Iꝝ Octobre take the ſame: foꝛ they beꝝ ſpecyall foꝛ the tꝛo꙯
ught all tymes of the peꝛe.From Apꝛill tyll Septembre ȳ tꝛo꙯
ugh leppyth.thenne angle to hym wyth a dubbyd hoke acoꝛdyn
ge to the moneth)Whyche dubbyd hokys pe ſhallſynde iꝝ then
de of this tꝛeatyſe; and the monethys wyth theym.:

℧ The grapllynge by a nother name callyd vmbre ia a de꙯
lycyous fyſſhe to mannys mouthe . And pe maye take
hym lyke as pe doo the tꝛought. And thyſe beꝝ his bay꙯
tes.℧ Iꝝ Marche ⁊ in Apꝛyll the redde woꝛme.℧ Iꝝ May the
grene woꝛme: a lytyll breyled woꝛme: the docke canker.and the
hawthoꝛꝝ woꝛme ∙ ℧ Iꝝ Iune the bayte that bredyth betwene
the tꝛee ⁊ the barke of aꝝ oke.℧ IꝝIulll a bayte that bredyth
oꝝ a ferꝝ lepf: and the grete redde woꝛme. And nyppe of the he
de: and put oꝝ pour hoke a codwoꝛme befoꝛe..℧ Iꝝ Auguſt the
redde woꝛme: and a docke woꝛme. And al the peꝛe after.a redde
woꝛme.

℧ The barbyll is a ſwete fyſſhe: but it is a quaſy meete ⁊ a
peꝛpllous foꝛ mannys body . Foꝛ compnly he peꝛipth
aꝝ intꝛoduꝛioꝝ to ȳ ſebrꝫs.Aud yf he be eteꝝ rawe: he
maye be cauſe of mannys dethe:whyche hath oft be ſeeꝝ Thy꙯

se be his baptes. ¶Iŋ Marche ⁊ iŋ Aprpll take fayr freſſhe chese : and laye it oŋ a boɜde ⁊ kptte it iŋ ſmall ſquare pecps of the lengthe of pour hoke. Take thenne a candpl ⁊ brenne it oŋ the ende at the popnt of pour hoke tpll it be pelow. And theŋe bpnde it oŋ pour hoke with fletchers ſplke : and make it rough lpke a welbede. This bapte is good all the ſomer ſeaſoŋ. ¶Iŋ Map ⁊ June take ẏ hawthoɜŋ woɜme ⁊ the grete redde woɜme. ⁊ŋd npppe of the heed. And put oŋ pour hoke a codwoɜme before. ⁊ that is a good bapte. Iŋ Jupll take the redde woɜme foɜ chepf ⁊ the hawthoɜŋ woɜme togph. Alſo the water docke lepf woɜme ⁊ the hoɜnet woɜme togpder. ¶Iŋ Auguſt ⁊ foɜ all the pere take the talowe of a ſhepe ⁊ ſofte cheſe : of eche plpke moche : and a lptpll honp ⁊ grpnde oɜ ſtampe thepm togph longe. and tempɜe it tpll it be tough. And put therto floure a lptpll ⁊ make it oŋ ſmalle pellettps. And ẏ is a good bapte to angle wpth at the grounde And loke that it ſpnke iŋ the water. oɜ ellps it is not good to this purpoos.

He carpe is a depntous fpſſhe : but there beŋ but felbe iŋ Englonde. And therfoɜe J wrpte the laſſe of hpm. He is aŋ eupll fpſſhe to take. Foɜ he is ſoo ſtɜonge enarmpd iŋ the mouthe that there mape noo weke harnapsholde hpm. And as touchpnge his baptes J haue but lptpll knowlege of it And me were loth to wrpte moɜe thaŋ J knowe ⁊ haue proupd But well J wote that the redde woɜme ⁊ the menow beŋ good baptps foɜ ḣpm at all tpmes as J haue herde ſape of perſones ⁊edpple ⁊ alſo founde wrpteŋ iŋ bokes of credence.

The cheupſh is a ſtatelp fpſſhe ⁊ his heed is a depty moɜſell. There is noo fpſſhe ſoo ſtɜonglp enarmpd wpth ſcalps oŋ the bodp. And bi cauſe he is a ſtɜonge bpter he ha the the moɜe baptesſwhiche beŋ thpſe. ¶Iŋ Marche the redde woɜme at the grounde : Foɜ complp thenne he woll bpte there at all tpmes of ẏ pere pf he be onp thinge hungrp. ¶Iŋ Aprpll the dpche canker that bredith iŋ the tɜee. A woɜme that bredith betbene the rpnde ⁊ the tɜee of aŋ oke The redde woɜme : and the ponge froſhps whaŋ the fete beŋ kpt of. Alſo the ſtone flpe the bobbe vnder the cowe toɜde : the redde ſnaplle. ¶Iŋ Map ẏ

i j

bapte that bredpth oŋ the ofper lepf ⁊ the docke canker togpꝰ
vpoŋ pour hoke. Alſo a bapte that bredpth oŋ a ferŋ lepf: ŷ cod
worme. and a bapte that bredpch oŋ aŋ halbthorŋ. And a bapte
thar bredpth oŋ aŋ oke lepf ⁊ a ſplke worme ⁊ a codworme to﹦
gpder. ¶ Iŋ June take the creket ⁊ the dorre ⁊ alſo a red wor﹦
me: the heed hptte of ⁊ a codworme befoze: and put thepm oŋ ŷ
hoke. Alſo a bapte iŋ the ofper lepf: ponge froſhps the thre-fete
hitte of bp the bodp: ⁊ the fourth bp the knee. The bapte oŋ the
halbthorŋ ⁊ the codworme togpder ⁊ a grubbe that bredpth iŋ
a dunghpll: and a grete greſhop. ¶ Iŋ Iupll the greſhop ⁊ the
humbplbee iŋ the medolb. Alſo ponge bees ⁊ ponge hornettes.
Alſo a grete brended flpe that bredith iŋ pathes of medolbes ⁊
the flpe that is amonge ppſmeers hpllps . ¶ Iŋ Auguſt take
wortwormes ⁊ magotes vnto Mpghelmas. ¶ Iŋ Septembre
the redde worme: ⁊ alſo take the baptes whaŋ pe map gete the
pm: that is to wpref Cherpes: ponge mpce not heerpd: ⁊ the hou
le combe.

The breeme is a noble tpſhe ⁊ a depntous. And pe ſhall
angle for hpm from Marche vnto Auguſt wpth a redde
worme: ⁊ theñe wpth a butter flpe ⁊ a grene flpe. ⁊ Wirh
a bapte that bredpth amonge grene rede: and a bapte that bre
dpth iŋ the barke of a deed tree. ¶ And for bremettis: take mag
gotes. ¶ And fro that tpme forth all the pere after take the red
worme: and iŋ the rpuer brolbne breede. Moo baptes theze beŋ
but thep beŋ not eaſp ⁊ therfore I lete thepm paſſe ouer.

A Tenche is a good fpſh: and heelith all maneꝛe of other
fpſhe that beŋ hurte pf thep mape come to hpm . He is
the moſt parte of the pere iŋ the mudde. And he ſtprpth
mooſt iŋ June ⁊ Iulp: and iŋ other ſeaſons but lptpll. He is aŋ
eupll bpter. his baptes beŋ thpſe. For all the pere brolbne bree
de toſtpd wpth honp iŋ lpknefſe of a butterpd loof: and the gre
te redde worme. And as for chepf take the blacke blood iŋ ŷ her
te of a ſhepe ⁊ floure and honp. And tempre thepm all togpder
ſomdeale ſofter than paaſt: ⁊ anopnt therlbpth the redde wor﹦
me: bothe for this fpſhe ⁊ for other. And thep woll bpte moche
the better therat at all tpmes.
¶ The percche is a dapnteuous fpſhe ⁊ paſſpnge hollom and

a free bytyng. Thise ben his baytes. In Marche the redde wor
me. In Aprill the bobbe vnder the cowe torde. In May the slo
thorn worme & the codworme. In June the bayte that bredith
in an olde fallen oke & the grete canker. In Jupll the bayte that
bredyth on the osyer lefe & the bobbe that bredeth on the dung
hyll: and the halwthorn worme & the codworme. In August the
redde worme & maggotes. All the yere after the red worme as
for the beste.

¶ The roche is an easy fysshe to take: And yf he be fatte & pen
nyd thenne is he good meete. & thyse ben his baytes. In Mar
che the most redy bayte is the red worme. In Apryll the bobbe
vnder the cowe torde. In May the bayte p̄ bredyth on the oke
lefe & the bobbe in the dunghyll.   In June the bayte that bre
dith on the osyer &, the codworme.   In Jupll hous flyes. & the
bayte that bredith on an oke. and the not worme & mathewes &
maggotes tyll Myghelmas. And after p̄ the fatte of bakon.

¶ The dace is a gentyll fysshe to take. & yf it be well refet then
is it good meete. In Marche his bayte is a redde worme. In
Apryll the bobbe vnder the cowe torde. In May the docke can
ker & the bayte on y slothorn and on the oken lefe. In June the
codworme & the bayte on the osyer and the whyte grubbe in y
dunghyll. In Jupll take hous flyes & flyes that brede in ppī
mer hylles: the codworme & maggotes vnto Mighelmas. And
yf the water be clere pe shall take fysshe whan other take none
And fro that tyme forth doo as pe do for the roche. For comyn
ly theyr bytynge & theyr baytes ben lyke.

¶ The bleke is but a feble fysshe. yet he is holsom   Hys baytes
from Marche to Myghelmas be the same that I haue wryten
before. For the roche & darse saupnge all the somer season asmo
che as pe maye angle for hym wyth an house flye : & in wynter
season w bakon & other bayte made as pe herafter may know.

¶ The ruf is ryght an holsom fysshe: And pe shall angle to hym
wyth the same baytes in al seasons of the yere & in the same wi
se as I haue tolde you of the perche: for they ben lyke in fysshe
& fedinges saupnge the ruf is lesse. And therfore he must haue y
smaller bayte.

¶ The flounder is an holsom fisshe & a free. and a subtyll byter
in his manere : For comynly whan he soukyth his meete he fe=

opth at grounde. ⁊ therfore þe must angle to hym wyth a gro
unde lyne lyenge. And he hath but one manere of bapte. ⁊ that
is a red worme. whiche is moost chepf for all manere of fysshe.
¶The gogen is a good fisshe of the mocheneſ:⁊ he bpteth wel
at the grounde. And his baptes for all the pere ben thple.ŷ red
worme: codworme: ⁊ maggotes.And þe must angle to him wt
a flote.⁊ lete pour bapte be nere ŷ botom or ellis on ŷ groñde.
¶The menow whan he shpnith in the water theñ is he bpttpr
And though his body be lptpll pet he is a rauenous biter ⁊ an
egre. And þe shall angle to hym wyth the same baptes that þe
doo for the gogpn:sauynge they must be smalle.
¶The ele is a quasp fysshe a rauendur ⁊ a deuourer of the bro
de of fysshe.And for the ppke also is a deuourer of fysshe I put
them bothe behpnde all other to angle.For this ele þe shall fpn
de an hole in the grounde of the water. ⁊ it is blewe blackpsshe.
there pur in pour hoke tpll that it be a fote wpthin ŷ hole.and
pour bapte shall be a grete angpll twptch or a menow.
¶The ppke is a good fysshe:but for he deuourpth so manp as
well of his owne kpnde as of other:I loue hpm the lesse. ⁊ for
to take hpm þe shall doo thus.Take a codlpnge hoke :⁊ take a
roche or a fresshe heering ⁊ a wpre wpth an hole in the ende:⁊
put it in at the mouth ⁊ out at the taplle downe bp the rpdge of
the fresshe heerpng.And thenne put the lpne of pour hoke in af
ter.⁊ drawe the hoke in to the cheke of ŷ fresshe heerpng.Theñ
put a plumbe of lede vpon pour lpne a perde longe from poure
hoke ⁊ a flote in mpdwaye betwene:⁊ caste it in a ppcte where
the ppke vspth.And this is the beste ⁊ moost surest rafte of ta
kpnge the ppke. ¶A nother manere takpnge of hpm there is.
Take a frosshe ⁊ put it on pour hoke at the necke bptwene the
shpnne ⁊ the body on ŷ backe half. ⁊ put on a flote a perde ther
fro:⁊ caste it where the ppke hauntpth and þe shall haue hpm.
¶A nother manere.Take the same bapte ⁊ put it in Asa fetida
⁊ rast it in the water wpth scorde ⁊ a corke: ⁊ þe shallnot fapll
of hpm.And pf þe lpst to haue a good sporte:thenne tpe the cor
de to a gose fote:⁊ þe shall se god halpnge whether the gose or
the ppke shall haue the better.
NOw þe wote with what baptes ⁊ how þe shall angle to
euerp manere fysshe. Now I woll tell pou how þe shall

kepe and fede your quycke baytes. ye ſhall fede and kepe theṁ
all in generall:but euery manere by hymſelf wyth ſuche thyngꝭ
in and on whiche they brede.And as longe as they ben quycke
⁊ newe they ben fyne . But. whan they benlyn a ſlough oꝛ elles
deed thenne ben they nought.Oute of thyſe ben excepted thre
brodes:That is to wyee of hoꝛnettys:humbylbees.⁊ waſpps.
whom ye ſhall bake in breede ⁊ after dyppe theyr heedes in blo
de ⁊ lete them dꝛye.Alſo excepte maggotes:whyche whan they
ben bredde grete wyth theyr naturell fedynge:ye ſhall fede the
pṁ ferthermoꝛe wyth ſhepes talow ⁊ wyth a cake made of flou
re ⁊ hony.thenne woll they be moꝛe grete . And whan ye haue
clenſyd theṁ wyth ſonde in a bagge of blanket kepte hote vn
der pour gowne oꝛ other warṁ thyng two houres oꝛ thre.theṁ
ben they beſte ⁊ redy to angle wyth. And of the froſſhe kytte þ
legge by the knee.of the graſſhop the leggys ⁊ wynges by the
body.

¶Thyſe ben baytes made to laſte all the yere.fyptſe ben flou
re ⁊ lene fleſſhe of the hepis of a conp oꝛ of a catte:virgyn wex
⁊ ſhepps talowe:and braye theṁ in a moꝛter:And thenne tem
pre it at the fyre wyth a lytyll purpfyed hony:⁊ ſoo make it vp
in lytyll ballys ⁊ bayte ther wyth your hokys after theyr quan
tyte.⁊ this is a good bayte foꝛ all manere freſſhe fyſſhe.

¶A nother,take the ſewet of a ſhepe ⁊ cheſe in lyke quantyte:⁊
braye theim togider longe in a moꝛtere:And take thenne floure
⁊ tempre it ther wyth.and after that alaye it wyth hony ⁊ ma=
ke ballys therof.and that is foꝛ the barbyll in eſpecyall.

¶A nother foꝛ darſe.⁊ roche ⁊ bleke:take whete ⁊ ſethe it well
⁊ thenne put it in blood all a daye ⁊ a nyghte.and it is a good
bayte.

¶Foꝛ baytes foꝛ grete fyſſh kepe ſpecyally this rule.whan ye
haue take a grete fyſſhe:vndo the mawe.⁊ what ye fynde ther=
in make that your bayte:foꝛ it is beſte.

¶Thyſe ben the.xh.flyes wyth whyche ye ſhall angle to þ tro
ught ⁊ graplyng:and dubbe lyke as ye ſhall now here me tell.

¶Marche.

The donne flpe the body of the donne woll ⁊ the wyngis of the pertryche. A nother doone flpe.the body of blacke woll:the wynges of the blackyst drake:and the Iap vnd the wpnge ⁊ vnder the taple.   ❡Aprpll.

❡The stone flpe.the body of blacke wull : ⁊ pelowe vnder the wpnge.and vnder the taple ⁊ the wpnges of the drake. In the begynnynge of May a good flpe.the body of roddyd wull and lappid abowte wpth blacke splke: the wpnges of the drake ⁊ of the redde capons hahpll.   ❡May.

❡The pelow flpe.the body of pelow wull : the wpnges of the redde cocke hahpll ⁊ of the drake lpttyd pelow.The blacke lou per.the body of blacke wull ⁊ lappyd abowte wpth the herle of y̵ pecok taple:⁊ the wpnges of y redde capon w̵ a blewe heed.

❡Iune.   ❡The donne cutte:the body of blacke wull ⁊ a pe = low lpste after epther spde : the wpnges of the bosarde bounde on wyth barkyd hempe.The maure flpe.the body of doske wull the wpnges of the blackest maple of the wylde drake.The tan dp flpe at sapnt Wpllpams dape. the body of tandy wull ⁊ the wpnges contrarp epther apenst other of the whitest maple of y̵ wylde drake.   ❡Iupll.

❡The waspe flpe.the body of blacke wull ⁊ lappid abowte w̵ pelow threde:the winges of the bosarde.The shell flpe at sapnt Thomas dape. the body of grene wull ⁊ lappyd abowte wpth the herle of the pecoks taple:wpnges of the bosarde.

❡August.   ❡The drake flpe.the body of blacke wull ⁊ lap= ppd abowte wpth blacke splke:wpnges of the maple of the blac ke drake wpth a blacke heed.

❡Thyse fpgures are put here in ensample of pour hokes.

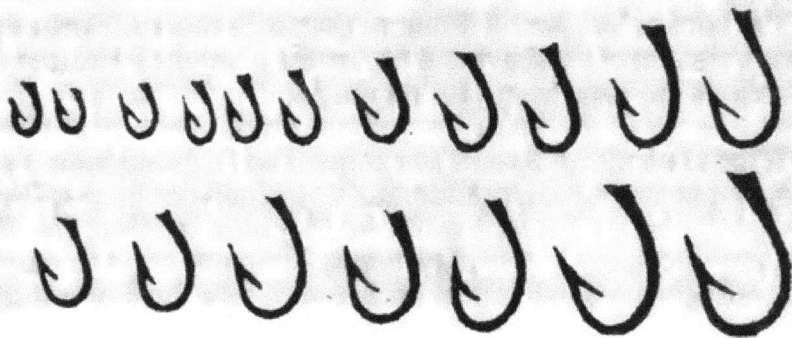

¶ Here folowpth the order made to all those whiche shall haue the vnderstondynge of this forsayde treatyse ẽ vse it for thepr pleasures.

HE that can angle ẽ take fysshe to pour plesures as this forsayd treatyse techpth ẽ shewpth pou: I charge ẽ requpre pou in the name of alle noble men that pe fysshe not in noo poore mannes feuerall water:as his ponde:ftewe:or other necessarp thynges to kepe fysshe in wpthout his lpcence ẽ good wpll. ¶ Nor that pe vse not to breke noo mannps gpnnps lpenge in thepr weares ẽ in other places due vnto thepm. Ne to take the fysshe awape that is taken in thepm.For after a fysshe is taken in a mannps gpnne pf the gpnne be laped in the compn waters:or elles in fuche waters as he hiteth;it is his owne propre goodes.And pf pe take it awape pe robbe hpm:whpche is a rpght shamfull bede to onp noble man to do p that the ups ẽ brpbours done:whpche are punpsshed for thepr eupll dedes bp the necke ẽ otherwpse whan thep mape be afpped ẽ taken.And also pf pe doo in lpke manere as this treatise shewpth pou:pe shal haue no nede to take of other mefips:whiles pe shal haue pnough of pour owne takpng pf pe lpste to labour therfore.whpche shall be to pou a verp pleasure to fe the fapr brpght shpnpnge fcalpd fysshes dpfcepued bp pour craftp meanes and drawen vpon londe. ¶ Also that pe breke noo mannps heggps in gopnge abowte pour dpfportes:ne oppn noo mannes gates but that pe shptte thepm agapn. ¶ Also pe shall not vfe this forsapd craftp dpfporte for no couetpfenes to thencreafpnge ẽ fparpnge of pour monep oonlp;but prpncppallp for pour folace ẽ to caufe the helthe of pour bodp.and fpecpallp of pour foule . For whanne pe purpoos to goo on pour difportes in fysshpng pe woll not defpre gretlp manp perfones wpth pou.whiche mpghte lette pou of poun game.And thenne pe mape ferue god deuowtlp in fapenge affectuouflp poure cuftumable praper. And thus dopnge pe shall efchewe ẽ vopde manp vices.as pdplnes whpche is prpncppall caufe to enduce man to manp other vp = ces.as it is rpght well knowen. ¶ Also pe shall not be to rauenous in takpng of pour fapd game os to moche at one tpme:whiche pe mape lpghtlp doo pf pe doo in euerp popnt as this prefent treatpfe shewpth pou in euerp popnt.whpche sholde lpght

ly be occalyon to byſhope your oWne byſportes ⁊ other men=
nys alſo. As Whan ye haue a ſuffyryent meſe ye ſholde couepte
nomore as at that tyme. ⸿Alſo ye ſhall beſye pourſelfe to nou=
ryſſh the game in all that ye maye; ⁊ to byſhope all ſuche thyn
ges as ben deuourers of it. ⸿And all thoſe that done after this
rule ſhall haue the bleſſynge of god ⁊ ſaynt Petyr; Whyche he
theym graunte that Wyth his precyous blood vs boughte.

⸿And for by cauſe that this preſent treatyſe ſholde not come
to the honops of eche pole perſone Whyche Wolde deſire it yf it
Were enprynted allone by it ſelf ⁊ put in a lytyll plaunflet ther
fore J haue complylyd it in a greter volume of dyuerſe bokys
concernynge to gentyll ⁊ noble men to the entent that the for
ſaph pole perſones Whyche ſholde haue but lytyll meſure in the
ſaph byſporte of fyſſhyng ſholde not by this meane vtterly byſ
trope it.

www.ingramcontent.com/pod-product-compliance
Lightning Source LLC
Chambersburg PA
CBHW032143080426
42733CB00008B/1186